REPORT

OF THE

DIRECTORS

OF THE

Northampton Coal Company,

CONTAINING A

REPORT OF THE ENGINEER

AND

BY-LAWS OF THE COMPANY.

Easton, Pa.
PRINTED BY JOSIAH P. HETRICH
1855.

DIRECTORS.

DAVID WEAVER,
C. R. HŒBER,
JOHN W. LESCHER,
THOMAS BARR,
DANIEL SEIGFRIED,
SAMUEL WEAVER,
JOHN LAUBACH, (MILLER.)

PRESIDENT.

DAVID WEAVER.

SECRETARY AND TREASURER.

JACOB FATZINGER.

REPORT.

To the Stockholders of the Northampton Coal Company.

The Board of Directors while publishing the By-Laws of said Company, and the report of the Engineer, desire to make a brief statement in reference to the location, &c., of the land belonging to said Company. The Real Estate which contains seven hundred and thirty acres of land, is located in Newport township, Luzerne county, Pa., about eleven miles below Wilkesbarre, in the great coal basin of the Wyoming Valley. According to the report of our able Engineer, Wm. J. Harlan, Esq., of Mauch Chunk, which is hereunto annexed, giving an estimate of the quantity of coal, &c.—two-thirds of the land contains White Ash Anthracite coal of superior quality. For the amount of coal per acre, the contemplated Rail Road, the facilities for getting coal to the different markets, we refer to said report. The company is organized under the General Mining and Manufacturing Law passed in 1849, and its subsequent amendments. The Capital Stock is now four hundred thousand dollars, and under the law, it may be increased to five hundred thousand dollars. The companies organized under this law, are at liberty to declare as large dividends to their Stockholders as the net profits earned will enable them to do. Companies under this law may hold two thousand acres of land, which enables us to extend our real estate one thousand two hundred and seventy acres, if to the interest of the company so to do. The individual liability of the Stockholders is much less than in Banks, being limited to debts for wages of laborers employed by them, for machinery, provisions, merchandize, country produce, and materials furnished to such company. Over three hundred thousand dollars of the capital stock is already sold, the remainder is now in market and can be had by applying to either of the Directors, or of the Stockholders. It is hoped that coal can be offered in market by the first day of August, A. D., 1856; we have expended about three hundred dollars in opening the mines. The company was organized on the 15th day of August, 1854. Wm. J. Harlan, Esq., was employed as Engineer, to explore the region and ascertain the most practicable route for a Rail Road, and make a geological examination of the property. The result of his labors is the annexed report. The By-Laws were adopted by a meeting of the Stockholders for the future government of the company.

By order of the Board.

DAVID WEAVER, *President.*

Weaversville, April 25, 1855.

THE ENGINEER'S REPORT

To the President and Directors of the Northampton Coal Company.

GENTLEMEN:—Having carefully examined your property, situated in Newport township, Luzerne county, Pa. my opinion is that about two-thirds of the whole number of acres, equal to five hundred acres, is underlayed with a superior quality of White Ash Anthracite Coal.

Assuming that my estimate is correct and taking

the aggregate thickness of the three veins at twenty eight feet, the property would yield 45,000 tons per acre, or 22,500,000 in the aggregate.

Of this quantity, one quarter for pillars would be required to support the mines, together with refuse and dirt, would reduce the *actual* production of *merchantable* coal to 30,000 tons per acre, or 15,000,000 tons in the aggregate, *if no fault occurs!*

The coal, in the ground, with a rail road to bring it into market, would be worth at *least* twenty-five cents per ton, yielding an aggregate sum of $3,750,-000! Your property has considerable timber upon it, which will be valuable for building purposes, and for the use of the mines.

Three-fourths of all your coal land lies on the summit of the Valley, between the "Hogback Ridge" and the Wilkesbarre or South Mountain. This portion must be tapped by a Rail Road leading either to Shickshinny or to Nanticoke upon the "North Branch Canal," following the South Mountain. The residue lies in a valley between the "Hogback Ridge" and the Susquehannah Mountain. This portion cannot be tapped by your proposed rail road.

The surface of the ground on the summit is uneven and rolling, producing a succession of small valleys and projections, between the northern and southern limits of the coal measure. The veins crop out on the South Mountain, and could be intersected by a tunnel driven at right angles from the valley below, affording considerable working "breast" of coal above the water level.

The best method of working the property would be in connexion with the adjoining land laying east of your property, and leading to the head of the valley. At this point, one hundred and fifty feet lower, running gangways longitudinally into the veins. By this means, the lowest possible point above the water level is attained, abridging the length of your road, increasing the length of coal breast, avoiding a plane, and materially reducing the cost of opening and working your mines.

The distrance from the western line of the "Davis" tract to the Susquehannah River at Shickshinny, is four and one half miles—from the same point to the Susquehannah River at Nanticoke, is five miles. If your property is opened at the *head* of the valley, the point indicated above, the distance will be four miles and a half, over favorable ground.

The route to Shickshinny is unfavorable, the descent is in rapids at successive intervals, making it necessary to change the grades frequently, and follow the side of the mountain. The route to Nanticoke is preferable, avoiding an inclined plane and reducing the elevation and grades to uniformity,

The property lies three hundred and ninety-eight feet above the Susquehannah River at Shickshinny, and three hundred and seventy-eight feet above the Susquehannah at Nanticoke.

The difference in the cost of a Railroad between the two routes is immaterial. The market in which the coal is sold, the bulk of traffic, the different roads would compete for, and the avoidance of planes, (which are always objectionable in their construction,) must decide the choice of location. I would recommend the route to Nanticoke or some available point above. The maximum grade from the summit of your property to the apex of the Shickshinny plane would be fifty-six feet per mile, one hundred and seventy-five feet would have to be overcome by a self-acting plane, one thousand two hundred feet in length, to reach the bank of the Susquehannah river at a proper grade for a bridge over that stream. A suitable bridge for a railroad overhead and wagon-way underneath, would cost at least $50,000.

The maximum grade from the *head* of the Valley to the Susquehanna river at or near Nanticoke would be fifty feet per mile, over ground underlayed with choice coal, affording easy curvatures and choice of good location. If your coal land is opened upon the Davis tract by a tunnel intersecting the vein at right angles, a road constructed with small T rail three quarters of a mile in length, will be required to lead to the "Screens and Breakers" for preparing the coal. From this latter point to the head of the valley a self-acting plane becomes unavoidable, overcoming an elevation of seventy-eight feet.

Inasmuch as the expense of the tunnel, construction of a small T railroad three quarters of a mile long, a self-acting plane to overcome seventy-eight feet fall, and much valuable time will be saved to you by opening at the head of the main road. I would recommend the necessary arrangements for its consummation.

In case such arrangement is feasible, the expense of opening your mine and constructing a railroad to Nanticoke would be materially reduced. That part of your projected road, leading from Col. Lee's saw-mill to the Susquehanna river, would be expensive, unless arrangements of a satisfactory nature can be effected with Col. Lee, to occupy the opposite side of the stream.

My estimate follows the rocky side. If it is not feasible to occupy this ground of Col. Lee's, I would recommend another outlet about a mile above this point, avoiding Nanticoke, and coming out upon the Susqnehanna river above the town.

I have made no estimate for the "right of way," presuming that no proprietor holding land to be occupied by your road would object to the release of sufficient ground for the construction of a first class Locomotive road, in consideration of the enhancement in value of his lands thereby. I herewith submit an opproximate estimate of the cost of a Railroad from your property to the Susquehanna river by both routes.

From the Summit to Shickshinny.

Cost of Grading, Masonry, &c.	$28,924,72
" " Iron Superstructure, &c.	42,747,72
	$71,672,44

From the Summit to, at or near Nanticoke.

Cost of Grading, Masonry, &c.	$32,699,10
" " Iron Superstructure, &c.	430,38,66
	$75,737,76

From the Head of the Valley to near Nanticoke.

Cost of Grading, Masonry, &c.	$28,301,60
" " Iron Superstructure, &c.	39,673,95
	$67,975,55

Annexed you will find an estimate of the expenses of opening your property and putting it in condition to produce 50,000 tons of coal per annum.

In doing the amount of work required so remote from a lumber region, while there is timber on your property to warrant the construction of a steam sawmill, becomes indispensably necessary, and is a valuable auxiliary at a coal mine.

Estimate for opening Mines, &c.

Cost of Steam Saw-Mill (20 Horse Power Engine,)	$4,000,00
75 yards Rock Tunnel at $45 00 per yard,	3,375,00
100 " Main Gangway, $5 00 "	500,00
25 Blocks Miners' Houses 16 x 32, two stories, $300,	7,500,00
1 Store House and Superintendent's Residence,	3,000,00
1 Locomotive House (Stone,)	500,00

1 Set Schutes, Pockets, Breaker Screens and 20 Horse Power Engine for Cleaning Coal,		6,000,00
2 Lumber Wagons,	$70	140,00
20 Good Mules,	$125	2.500,00
20 Set Harness for Mules,	$10	200,00
40 Mine Cars, 2 tons each,	$100	4,000,00
1 Stable for Team Mules,		600,00
1 Set Pockets, Schutes, wh'ge & fixtures, for loading Boats,		5.650,00
50 Main road Cars. 4 tons each	$200,	10,000,00
1 Locomotive Engine, first class, 13 tons,		7,650,00
1 Smith and Car repair Shop,		500,00
		$56.115,00

In the above estimate the timber taken from the property is rated at its market value. If your property is opened at the *head* of the Valley, instead of the *summit* the cost of the tunnel will be obviated, reducing the actual expenditure to $52,740,00.

The expense of mining, preparing and delivering your coal into Boats cannot exceed (if property managed) seventy-five cents per gross ton, allowing two cents per ton per mile for tolls and transportation upon your Road.

Coal delivered into Boats at Nanticoke is worth one dollar and a half per gross ton, leaving you seventy-five cents per ton nett profit on coal, which on a business of 50,000 tons per annum would yield $37,-500 nett profit per annum, equal to a dividend of *nine per cent.* on the Capital Stock.

When we take into consideration the connexions your road may make, by a Bridge near Nanticoke, crossing the Susquehanna river, with the "Scranton and Bloomsburg Road" in connexion, leading East, West, North and South, the large amount of coal land over which it must pass, inviting a large and profitable traffic thereon, the quantity of coal you possess, and the extent to which you may enlarge the production of your mines at a relative increased outlay, the growing interests and inviting markets, opening North and upon the Lakes ; your enterprise if properly managed, cannot fail to prove highly remunerative, yielding a handsome dividend on the outlay.—It is such Stock as is not to be found every day in market, and invites the attention of Capitalists as a permanent investment, not to be equalled by any stock now offering in the market.

Considering the increased consumption of coal, its utility and cheapness compared with wood, the inefficient means in operation to meet the demand of the markets *now* opened to its use ; it is a matter of surprise with practical men, that men who possess the means and ability have not put their means and efforts into active service to meet the growing wants of the trade and reap the rich harvest which awaits such an investment. All of which is resp'y submitted.

WM. J. HARLAN, *Engineer.*

Mauch Chunk, Sept. 10, 1854.

Articles of Association of the Northampton Coal Co.

OBJECT.

The Subscribers hereby associate themselves together for the purpose of mining Anthracite coal and other minerals, and transporting the same to market, under the certificate approved of by the Attorney General of the State of Pennsylvania, and filed in the Office of the Secretary of State at Harrisburg, which is in accordance with the act of the General Assembly of the Commonwealth of Pennsylvania, entitled "An Act to encourage Manufacturing Operations in this Commonwealth," approved April 7th, A. D., 1849, and the several supplements thereunto, and for themselves and their assigns they adopt the following as the articles of Association of the said Company.

ARTICLE I.

SECTION 1. The corporative name of said Company shall be "The Northampton Coal Company," their

Lands being located in the county of Luzerne, in the State of Pennsylvania.

SEC. 2. The amount of the Capital Stock of said Company shall be $400,000, divided into 8,000 shares of fifty dollars each.

SEC. 3. The existence of the said Company shall terminate twenty years from the day upon which the certificate required by law shall be filed in the office of the Secretary of the Commonwealth.

ARTICLE 2.

SEC. 1. The corporation may purchase, hold and convey Real Estate as follows, viz :—Such real estate as they may deem necessary for the convenient and successful prosecution of its mining operations not exceeding in all to 1000 acres.

SEC. 2. Such right of way as may be necessary for the construction of a rail road, or rail roads for the conveyance of the products of their mines to a convenient landing.

SEC. 3. To purchase or lease any real estate in a Lot or Lots not exceeding 5 acres for the purpose of storing coal at the place most convenient for shipping the same, and for putting up the necessary buildings, wharves, &c.

ARTICLE 3.

SEC. 1. The management of this company is vested in a Board of Managers or Directors, which shall consist of seven persons.

SEC. 2. The following persons shall be the Directors of this company, to act until their successors shall be elected at the annual meeting of the stockholders viz :—David Weaver, Daniel Siegfried, C. R. Hœber, J. J. Slocum, Jno. W. Lescher and Thomas Barr.

SEC. 3. The first annual meeting of the stockholders shall be held on the last Thursday of November next, when the first election of Directors to supercede those hereinbefore appointed shall be held, and the annual meeting and election of Directors shall be held on that day of each year thereafter, at such convenient place as the Board of Directors shall appoint. The said election shall be conducted by three Inspectors, to be chosen by the stockholders at each annual meeting. Neither of said Inspectors shall be a Director of said company. The seven persons having the greatest number of votes shall be Directors and they shall enter upon the duties of their office on the next Thursday following that of the election. In voting for Directors the stockholders shall designate on each ticket one as President. If an equal number of votes shall be cast for two or more persons, the existing Board of Directors shall by ballot determine which of the same shall be appointed. If any Director shall cease to hold stock, his office shall be vacant. Whenever any vacancy occurs in the Board of Directors, or in the office of President, it shall be filled for the residue of the unexpired term by such eligible person as the Board of Directors may elect by ballot. If there should at any time be a failure to elect Directors at the time hereinbefore specified, it shall be the duty of the Board to take immediate measures for an election within sixty days.

Sec. 4. Each stockholder to be entitled to as many votes as he may hold certificates, provided that no one person either in his own right or by proxy shall cast more than one quarter of the whole number of shares subscribed for.

Sec. 5. Each stockholder must have held his stock in his own name at least thirty days prior to the time of voting.

Sec. 6. No person can have the right of voting by proxy in any election for Directors unless he resides more than ten miles from the place of holding the election, except females.

Sec. 7. The Board of Directors shall at each annual

meeting make a full and detailed report in writing of all the business and transactions of the company during the year. They shall also appoint a Secretary and Treasurer, and a General Manager of the Company, and such other officers, clerks, agents and servants as may be necessary to conduct the business of the said Company, fix their salaries and change the same at pleasure.

ARTICLE 4.

Sec. 1. The President of the Company shall preside and may vote at the meetings of the Directors, at which he may be present. He shall have authority to collect, receive and sue for all monies or property due or belonging to said Company; to submit to arbitration any controversies in which it may be involved—to cancel and satisfy any Judgment, Mortgage or Deed which it may hold, and to release or discharge the whole or any part of the property mortgaged or encumbered—to sell or transfer any stocks or other property belonging to, or pledged to the Company, and to receive any dividends accruing therefrom, under the direction of the Board of Directors and he may appoint for any of these purposes one or more Attorneys at Law or in fact, under him. In case of his inability to act, for any cause, the Board of Directors may authorize their Treasurer, principal Manager, or one of their own body to exercise any or all the powers of the President Pro Tempore.

Sec. 2. At each annual meeting, the Treasurer shall submit a detailed report in writing of all the receipts and disbursements of the Company.

Sec. 3. The President and Secretary shall sign all contracts made by the Company, evidence of debt and certificates of stock issued by it, and no other officer, director or stockholder, unless authorized by the Board of Directors, shall make any contract or engagement in its behalf.

Sec. 4. Minutes of the Board of Directors shall be kept, signed by a member of the Board appointed by them as clerk.

ARTICLE 5.

Sec. 1. The Board of Directors shall cause suitable certificates of stock to be prepared, and also suitable books to be kept for the registry and transfer of the shares of the Company, and every transfer shall be made on such books and signed by the shareholder, or his attorney duly authorized in writing.

Sec. 2. Every transfer shall be made and taken subject to all the conditions and stipulations contained in these articles.

Sec. 3. The Board of Directors may close the transfer books from time to time, as the convenience of the Company may require.

ARTICLE 6.

Sec. 1. Out of the funds of the Company, the Board of Directors shall defray its expenses, and twice in each year, may declare and pay to the stockholders, or their attorneys, such dividends from the net profits of its business, as they may deem expedient, notice of the declaration of dividends shall be published within ten days after declaring the same, in at least one weekly newspaper printed in Luzerne county, and one English and one German weekly paper in Northampton county. Interest on all unclaimed dividends shall belong to the Company.

Sec. 2. The Board of Directors may at any time propose amendments to these articles, which, when adopted by a majority of the stockholders at any regular meeting, shall become a portion of these articles.

Sec. 3. The President may call special meetings of the stockholders, at any time and place he may deem advisable, but on a written request signed by ten stockholders, or more, he shall call a meeting of the Company within twenty-five days of such request.

C

Summary of the Liabilities and Assets of the Lehigh Coal and Navigation Company—Jan. 1, 1848, *to Jan.* 1, 1855.

LIABILITIES.	Jan. 1, 1848.	Jan. 1, 1849.	Jan. 1, 1850.	Jan. 1, 1851.	Jan. 1, 1852.	Jan. 1, 1853.	Jan. 1, 1854.	Jan. 1, 1855.
Capital Stock,	1,503,550 00	1,503,550 00	1,503,550 00	1,503,550 00	1,503,550 00	1,569,800 00	1,985,850 00	2,474,600 00
Common Loans,	3,688,235 34	3,686,856 93	3,685,133 83	3,688,331 85	4,346,077 86	4,353,827 12	4,050,553 89	3,586,078 10
Mortgage Loan,	946,808 56	1,000,000 00	1,000,000 00	1,000,000 00	1,000,000 00	929,342 00	837,492 00	832,792 00
Floating Debt,	340,075 36	446,982 30	239,046 68	383,765 43	588,833 39	364,933 10	332,049 10	373,150 21
Unpaid Dividends,	798 70	798 70	798 70	798 70	798 70	3,823 70	3,503 95	4,245 70
Arrears of Interest,	867,095 48	873,941 27	824,174 65	639,874 81	99,085 39	74,812 89	67,283 71	58,634 59
	7,346,563 44	7,512,129 20	7,252,703 86	7,216,320 79	7,538,345 34	7,296,538 81	7,276,732 65	7,329,500 60
Balance to the credit of Profit and Loss account,	291,633 86	367,691 18	565,671 35	706,172 96	777,337 24	907,995 25	1,164,672 36	1,689,954 66 *
	7,638,197 30	7,879,820 38	7,818,375 21	7,922,493 75	8,315,682 58	8,204,534 06	8,441,405 01	9,019,455 26
ASSETS.								
Canal and River Improvements,	4,455,000 00	4,455,000 00	4,455,000 00	4,455,000 00	4,455,000 00	4,455,000 00	4,455,000 00	4,455,000 00
Lehigh and Susquehanna Rail-road,	1,369,820 75	1,380,798 00	1,380,000 00	1,380,000 00	1,380,000 00	1,380,000 00	1,380,000 00	1,380,000 00
Real Estate, cost of coal mine lands and other lands, Rail-roads to the several mines of the Company, and other improvements, Wharves and Landings at Philadelphia, &c.	1,193,044 48	1,324,235 69	1,337,289 93	1,353,528 01	1,347,248 77	1,307,876 55	1,280,998 17	1,281,901 46
Moveable effects, Debts due the Company, Bonds and Mortgages, and other securities,	612,475 53	712,246 33	635,520 84	717,090 30	1,103,301 47	1,012,927 68	1,249,910 11	1,845,397 23
Cash on hand,	7,856 54	7,540 36	10,564 44	16,875 44	30,132 34	48,729 83	75,496 73	57,156 57
	7,638,197 30	7,879,820 38	7,818,375 21	7,922,493 75	8,315,682 58	8,204,534 06	8,441,405 01	9,019,455 26

* Of the balance which appears to the credit of the Profit and Loss account, about $1,100,000 is now, April 25, 1855, invested in the name of Trustees, as a contingent fund. Of this sum, over $500,000 is invested in the loans of the Company, and the rest in other loans and available securities, the whole being included among the assets. **Of the remainder of the balance standing to the credit of Profit and Loss, a large sum has been expended in improvements of the Company's Real Estate.**

ENGINEER'S REPORT.

Office of the Lehigh Coal and Navigation Company,
Mauch Chunk, Jan. 1st, 1855.

JAMES COX, Esq.
President Lehigh Coal and Nav. Co.

SIR,—The water was let into the Lower Section of the navigation on the 10th of March last, and into the Upper Section on the 25th of the same month; but, in consequence of the continuance of cold weather, and accumulation of ice in the pools of the Upper Section, boats did not reach White Haven before the 6th of April.

Shipments of coal commenced at this place on the 23d of March; but the boatmen being dissatisfied with the rates of freight offered by the operators, generally declined taking loading, until the middle of April, when shipments from the different landings became quite active.

Near the close of the month of April, a freshet occurred in the Delaware River, which overflowed the banks of the Delaware Canal, doing much injury, which required about one month to repair. During this time shipments were quite limited, confined mostly to the local trade of the Lehigh and Morris Canals. Nothing further occurred, with some minor exceptions, to interrupt the regular course of business until the close of the navigation by ice, which took place much earlier than usual.

Notwithstanding the late period at which shipments commenced, the long suspension of the Delaware Canal, and the early closing of the navigation, there where shipped from the Lehigh 1,246,418 tons of coal, an increase of 165,878 tons over 1853.

The raising and strengthening the banks and other works, preparatory to the admission of an increased depth of water, has been continued.

All the locks, aqueducts, and dams, have been raised; and there now remains to be raised the banks of the South Easton level, two short pieces in the levels below Allentown and Bethlehem, and the short levels of canal on the Upper Section, to complete the preparations for the admission of six feet depth of water *throughout the navigation.*

It is proposed to raise the levels below the Guard Lock, at Allentown and the South Easton level, to a sufficient height to admit of seven feet water. This will supercede the necessity for working the Guard Locks, whenever the water shall be less than one foot in depth on the combing of the dams. Contracts have been made to have these levels raised, and a towing path constructed on the opposite or berm side of the canal, to be completed early in the spring of 1855.

About four and a half miles of towing path on the berm side of the canal, have been completed and brought into use; and it is expected that several miles more will be completed by the time the navigation shall be opened in the spring. Lock No. 39, Lower Section, which was in the course of being raised at the date of my last annual report, has been completed, and has fully answered the purpose for which it was intended. One of the narrow locks on the Upper Section, between Mauch Chunk and Penn Haven, has been widened to twenty-two feet, leaving two more to be widened. Arrangements have been made to widen one of them during the present winter.

The lock at the entrance to the canal at this place, was originally built of wood, and has now become so much decayed as to render it necessary that it should be rebuilt. Arrangements have been made, not only to rebuild it, but to double its length, so as to admit of the passage of *four* boats of the present size, at a single lockage, and to have it completed by the time the navigation shall be open in the spring of 1855.

Some of the locks require relining, and the outlet locks into the pools on the Lower Section, to have drop gates substituted for the swing gates now in use; all of which it is intended shall be done during the present winter. All the other works are now in good working order.

. The towing path along the south side of pool No. 8, being occupied by the Lehigh Valley Rail Road, I would recommend building a bridge across from the south side of the pool, about one-fourth of a mile above the Guard Lock, to Smith's Island, construct a towing path along the south side of the island, about one mile in length to its head, and then a causeway across to intersect the old towing path on the north side of the pool; thereby avoiding the delays and perplexities consequent upon ferrying across the pool, as heretofore practised.

As the business of the Lower Section has become quite large, (nearly up to its capacity *with the present sized boats,*) and there being now considerable delay at some of the locks, I would suggest the propriety of making arrangements for increasing the facilities, by doubling the capacity of the locks, either by doubling their length, or by building another lock along side of the present one; beginning with those at which there is the greatest detention, and proceed, from year to year, until the whole shall be doubled.

The descending navigation, from Stoddartsville to White Haven, continued in its usual navigable order throughout the season of 1854, excepting the extremely low water during the summer. No expenditures are required, excepting for ordinary repairs.

There has been expended on the Lower Section, for all purposes, $108,553.25.

Of this sum, there was expended for raising the lock, raising and strengthening the banks, completing Lock No. 39, and materials for rebuiding Lock No. 1, - - - - - - -	$29,028 22
Land for the use of the navigation, - -	1,065 60
The balance at other work, chiefly ordinary repairs,	78,459 43
	$108,553 25

There has been expended on the Upper Section, $33,457.61.

Of this sum, there was expended in widening locks, raising dams, &c. &c. - - - -	$13,065 40
Descending Navigation, - - -	610 05
The balance at ordinary repairs, - - -	19,782 16
	$33,457 61

It will be seen, that the expenditures on the navigation are much larger than in 1853. This is mainly owing to the high price which had to be paid for labour and materials.

Lehigh and Susquehanna Rail Road.

This work continued in good running order throughout the season, and required nothing more than ordinary repairs; for which there has been expended $8,526.34.

Coal Improvements.

Descending rail road, from the Summit to Mauch Chunk, and schutes.

Schute No. 2, and the pocket used for shipping stove and chestnut coal, had become so much decayed as to require rebuilding, which was done, and the work brought into use at the opening of business in the spring. Another screen was added, and the pocket was enlarged to three times its former capacity, which has materially added to the expedition in loading coal at those schutes.

There has been no expenditure on the main road excepting for ordinary repairs.

It is proposed to renew the tracks on schute No. 3, during the present winter, and lay it with a T rail, and also add a set of pockets to schute No. 1; dispense with the use of a rolling screen altogether, and appropriate this schute exclusively to the shipment of coal from the Old Tunnel.

There has been expended on these works, for all purposes, $14,706.54.

Back Track.

The tracks on Inclined Plane No. 1, have been renewed with a T rail. Embankments have been built in place of the high trestle-work at Plane No. 2; also in place of the one near the Summit, leaving the one on Plane No. 2, to be built during the present winter of 1854–5.

Arrangements have been made to relay the tracks on No. 2, with a T rail. Contracts have been made for the iron, and it is now being delivered on the ground.

The engines at both of the planes require some repairs; otherwise the works are all in good working order.

There has been expended on this work, including materials for renewing the tracks on Plane No. 2, $15,303.06.

Panther Creek Rail Roads.

These roads, as well as the inclined planes, continued in good working order throughout the season.

The light plate rails, which were originally laid on the road through the valley, about three miles in length, have been removed and heavier ones substituted. Some of these roads require relaying, particularly the back track, leading from the summit to the valley.

As suitable timber for rail roads is becoming quite scarce and high in price, I would recommend relaying this track with a T rail of 30 or 40 lbs. to the yard.

The inclined planes are now working nearly up to their capacity, and when an accident happens, whether from breaking of a rope, or from cars getting off the track, the interruption is felt throughout the whole line of works. This evil can, in a measure, be remedied by constructing a duplicate track and hauling up machinery, so arranged that if an accident occurs on *one* side, it can be thrown out of gear, and the other put in gear in a few minutes. This arrangement will allow the engines to continue running while the accident is being righted. The present graded surface of the inclined planes is of sufficient width to admit of such an arrangement, without any further expense of grading. The cost will be about $7000 each plane.

There has been expended on these works in repairs, and materials for relaying the inclined planes, $22,730.64.

Tunnel No. 2.

The back track leading to tunnel No. 2 has been completed; the track on the inclined plane laid; the breakers and foundation of the machinery built; the engine and other machinery on the ground ready to be put up; but, in consequence of the scarcity and high price of labour of all kinds, it was deemed advisable to postpone putting them up until the spring, especially as all the work can be completed before the contractors can be ready to send forward coal.

There has been expended on these works up to the present time, $28,816.45.

But little progress has been made in driving forward tunnel No. 9, during the past year. The interruption caused by the quicksand coming in, has continued up to the present time, although an open cut has been made nearly across the valley. It is believed, however, that the difficulty is about mastered, and it is hoped the work will now advance without any further interruption.

There has been expended upon it during the past year, $9,952.45.

Old Tunnel.

The road leading to and from the old tunnel mines, a length of over two miles, was completed; and over 10,000 tons of coal passed over it before the close of the season. The cost was $10,863.57.

There has been shipped and sent to market,—

From the Summit mines,	- -	413,049 tons.
Room Run,	- - -	92,138
Old tunnel,	- - -	10,730
Greenwood,	- - -	28,659
		541,576 tons.

All of which is respectfully submitted.

E. A. DOUGLAS,
Sup't and Engineer.

SUPPLEMENTAL REPORT.

Office of the Lehigh Coal and Navigation Company,
Mauch Chunk, April 13, 1855.

James Cox, Esq.
President Lehigh Coal and Nav. Co.

Sir,—The Lower Section of the Navigation was ready for the reception of water about the middle of March, and the Upper Section on the 25th of the same month; but in consequence of the inclement weather and ice in the canal, boats did not pass through to this place before the 30th: and up to this time the pools above Rockport are closed with ice.

For the same reason, the coal improvements have been quite backward in getting into order. On the 4th inst. a few boats were loaded with coal, but up to the present time the shipments have been quite limited. It is presumed, however, that they will be largely increased in a few days.

Yours respectfully,

E. A. DOUGLAS.

www.ingramcontent.com/pod-product-compliance
Lightning Source LLC
LaVergne TN
LVHW011146110826
845150LV00008B/2549

* 9 7 8 1 4 1 8 1 9 2 2 7 3 *